Atlas of the Human Skeleton

Matt Hutchinson • Jon Mallatt • Elaine N. Marieb

Photographs by Ralph T. Hutchings

An imprint of Addison Wesley Longman, Inc.

San Francisco Boston New York
Capetown Hong Kong London Madrid Mexico City
Montreal Munich Paris Singapore Sydney Tokyo Toronto

Publisher: Daryl Fox
Senior Project Editor: Kay Ueno
Managing Editor: Wendy Earl
Production Manager: Janet Vail
Production Editor/Photo Coordinator: Kelly Murphy
Photographer: Ralph T. Hutchings
Cover Designer: F. tani Hasegawa
Compositor: Precision Graphics

ISBN 0-8053-4988-X
5 6 7 8 9 10—VHP—04 03 02

Copyright © 2001, Benjamin Cummings, an imprint of Addison Wesley Longman, Inc.
All rights reserved.
Printed in the United States of America. This publication is protected by Copyright and permission should be obtained from the publisher prior to any prohibited reproduction, storage in a retrieval system, or transmission in any form or by any means, electronic, mechanical, photocopying, recording, or likewise. For information regarding permission(s), write to: Permissions Department.

Preface

This *Atlas of the Human Skeleton* was the joint inspiration of Elaine N. Marieb and Benjamin Cummings Publisher Daryl Fox. Packaged with *Human Anatomy & Physiology,* Fifth Edition, and *Human Anatomy,* Third Edition, the full-color views of the human skeleton are presented with a degree of clarity and scale that could never be achieved within a textbook alone.

The skeletal views were carefully chosen and organized by Elaine Marieb. Matt Hutchinson of Washington State University took on the arduous task of labeling each structure. Jon Mallatt, co-author of the human anatomy text, scrutinized and approved the views, leaders, and labels through several rounds of page proof. Figure references to related atlas images can be found in the illustration figure legends of both the anatomy and the anatomy and physiology textbooks.

Ralph T. Hutchings, formerly with The College of Surgeons of England, photographed each of the bone structures found in this book. His reputation as an anatomical photographer preceded him, and we certainly were not disappointed—the quality of his work is here for all to see. We are most grateful to him for lending his expertise to this project, and for his good humor and ready willingness to meet our demands.

We are hopeful that the *Atlas of the Human Skeleton* proves to be a relevant addition to the Marieb and Marieb/Mallatt texts. Benjamin Cummings would welcome your comments and suggestions, which may be sent to the following address:

Kay Ueno
Anatomy & Physiology
Benjamin Cummings Science
1301 Sansome Street
San Francisco, CA 94111

Contents

Figure 1	Skull, anterior view	1
Figure 2	Skull, right external view of lateral surface	2
Figure 3	Skull, internal view of left lateral aspect	3
Figure 4	Skull, external view of base	4
Figure 5	Skull, internal view of base	5
Figure 6	Occipital bone, inferior external view	6
Figure 7	**Frontal bone**	**7**
	(a) anterior view	7
	(b) inferior surface	7
Figure 8	**Temporal bone**	**8**
	(a) right lateral surface	8
	(b) right medial view	9
Figure 9	**Sphenoid bone**	**10**
	(a) superior view	10
	(b) posterior view	10
	(c) anterior view	10
Figure 10	**Ethmoid bone**	**11**
	(a) left lateral surface	11
	(b) posterior view	11
	(c) anterior view	11
Figure 11	**Mandible**	**12**
	(a) right lateral view	12
	(b) right medial view	12
Figure 12	**Maxilla**	**13**
	(a) right lateral view	13
	(b) right medial view	13
Figure 13	**Palatine bone**	**14**
	(a) right lateral view	14
	(b) right posterior view	14
Figure 14	Bony orbit	15
Figure 15	Nasal cavity, left lateral wall	16
Figure 16	**Fetal skull**	**17**
	(a) anterior view	17
	(b) lateral view	17
Figure 17	**Articulated vertebral column**	**18**
	(a) right lateral view	18
	(b) posterior view	19
Figure 18	**Various views of vertebrae C_1 and C_2**	**20**
	(a) atlas, superior view	20
	(b) atlas, inferior view	20
	(c) axis, superior view	21
	(d) axis, inferior view	21
	(e) articulated atlas and axis, superior view	21
Figure 19	**Cervical vertebrae**	**22**
	(a) right lateral view of articulated cervical vertebrae	22
	(b) fifth (typical) cervical vertebra, superior view	22
	(c) fifth (typical) cervical vertebra, posterior view	22
	(d) fifth (typical) cervical vertebra, right lateral view	23
	(e) vertebra prominens (C_7), superior view	23
Figure 20	**Thoracic vertebrae**	**24**
	(a) articulated thoracic vertebrae, right lateral view	24
	(b) seventh (typical) thoracic vertebra, superior view	24
	(c) seventh (typical) thoracic vertebra, posterior view	24
	(d) comparison of T_1, T_7, and T_{12} in right lateral views	25
Figure 21	**Lumbar vertebrae**	**26**
	(a) articulated lumbar vertebrae and rib cage, right lateral view	26
	(b) second lumbar vertebra, superior view	26
	(c) second lumbar vertebra, posterior view	27
	(d) second lumbar vertebra, right lateral view	27
Figure 22	**Sacrum and coccyx**	**28**
	(a) posterior view	28
	(b) right lateral view	28
	(c) anterior view	29
Figure 23	**Bony thorax**	**30**
	(a) anterior view	30
	(b) posterior view	31
	(c) sternum, right lateral view	32
	(d) sternum, anterior view	32
	(e) typical rib, posterior	33
	(f) articulated typical rib and vertebra, superior view (left) and lateral view (right)	33
Figure 24	**Scapula and clavicle**	**34**
	(a) right scapula, anterior view	34
	(b) right scapula, posterior view	34
	(c) right scapula, lateral aspect	35
	(d) right clavicle, inferior view (top) and superior view (bottom)	35
	(e) articulated right clavicle and scapula, superior view	35
Figure 25	**Right humerus**	**36**
	(a) anterior view	36
	(b) posterior view	36
	(c) proximal end, anterior view	37
	(d) proximal end, posterior view	37
	(e) distal end, anterior view	37
Figure 26	**Right ulna and radius**	**38**
	(a) articulated right ulna and radius, anterior view	38
	(b) articulated right ulna and radius, posterior view	38
	(c) articulated right humerus, ulna, and radius, anterior view	38
	(d) articulated right humerus, ulna, and radius, posterior view	38
	(e) right ulna, proximal end: anterior, posterior, medial, and lateral views	39

Figure 27	**Bones of the right hand**	**40**	
	(a) lateral aspect	40	
	(b) dorsal aspect	40	
Figure 28	**Bones of the male pelvis**	**41**	
	(a) right hip bone, lateral view	41	
	(b) right hip bone, medial view	41	
	(c) articulated male pelvis, anterior view	42	
	(d) articulated male pelvis, posterior view	43	
Figure 29	**Right femur**	**44**	
	(a) anterior view	44	
	(b) posterior view	44	
	(c) proximal end, anterior view	45	
	(d) proximal end, posterior view	45	
	(e) proximal end, medial view	45	
	(f) distal end, anterior view	46	
	(g) distal end, posterior view	46	
	(h) articulated right femur and patella, inferior view with knee extended	47	
	(i) right patella, anterior surface	47	
	(j) articulated right femur and patella, inferior posterior view with knee flexed	48	
	(k) right patella, posterior surface	48	
Figure 30	**Right tibia and fibula**	**49**	
	(a) articulated right tibia and fibula, anterior view	49	
	(b) articulated right tibia and fibula, posterior view	49	
	(c) right tibia, proximal end, anterior view	50	
	(d) right tibia, proximal end, posterior view	50	
	(e) right tibia, proximal end, articular surface	51	
	(f) articulated right tibia and fibula, proximal end, posterior view	51	
	(g) articulated right tibia and fibula, distal end, posterior view	51	
	(h) right fibula, proximal end, anterior view	52	
	(i) right fibula, proximal end, posteromedial view	52	
	(j) right fibula, proximal end, medial view	52	
Figure 31	**Bones of the right ankle and foot**	**53**	
	(a) superior surface	53	
	(b) inferior (plantar) surface	53	
	(c) medial view	54	
	(d) lateral view	54	
	(e) right calcaneus, superior aspect	55	
	(f) right calcaneus, posterior aspect	55	
	(g) right talus, inferior view	55	

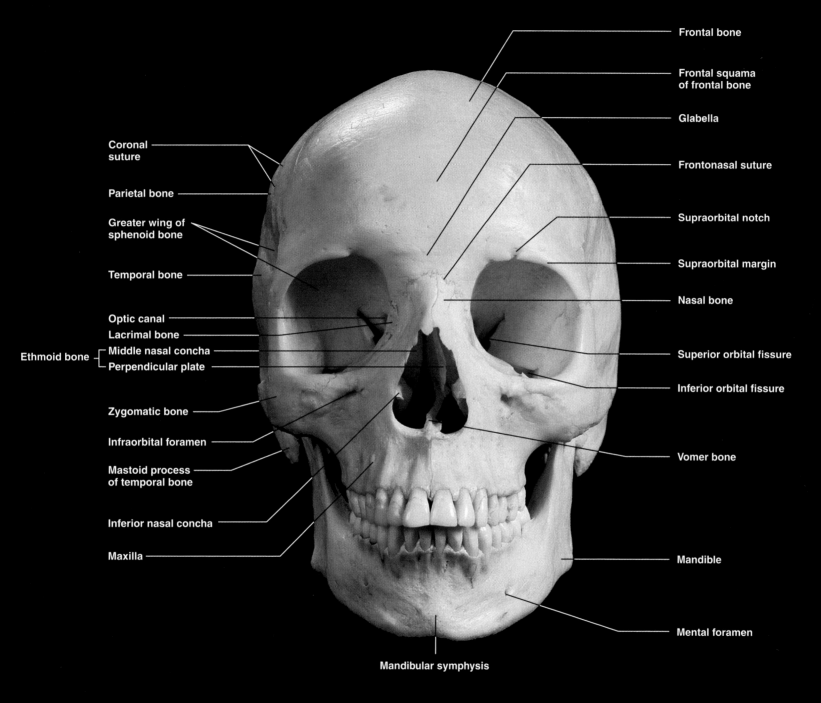

Figure 1 Skull, anterior view.

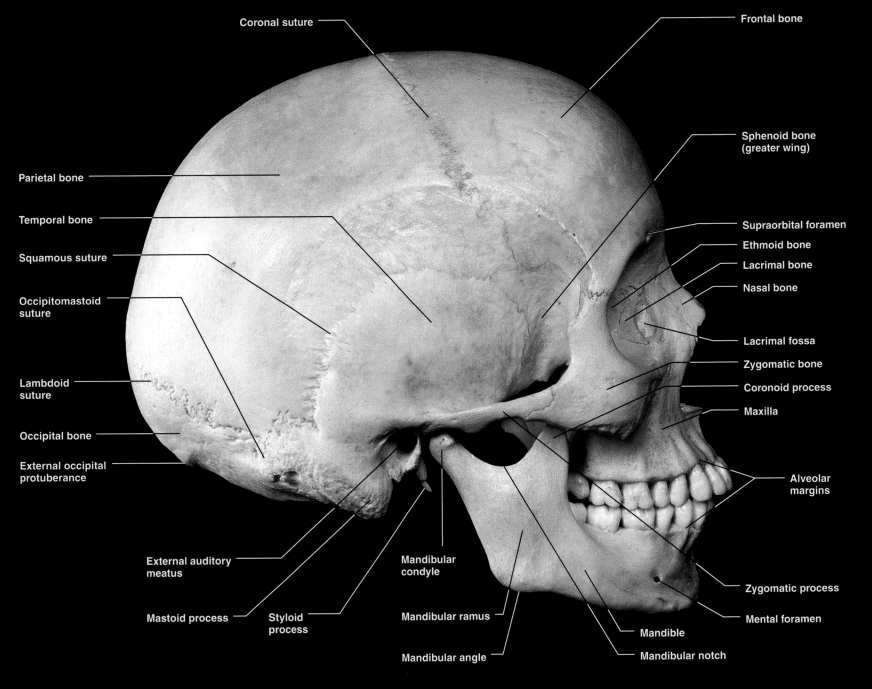

Figure 2 Skull, right external view of lateral surface.

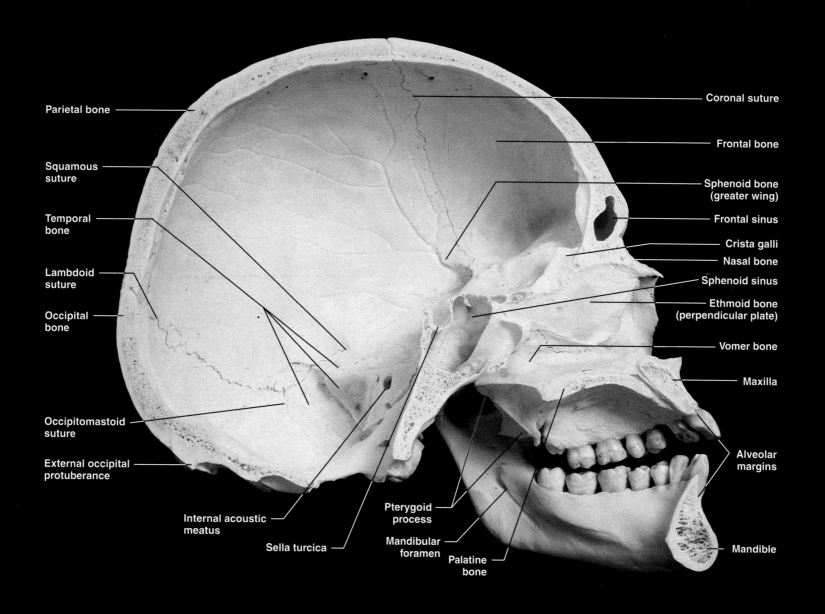

Figure 3 Skull, internal view of left lateral aspect.

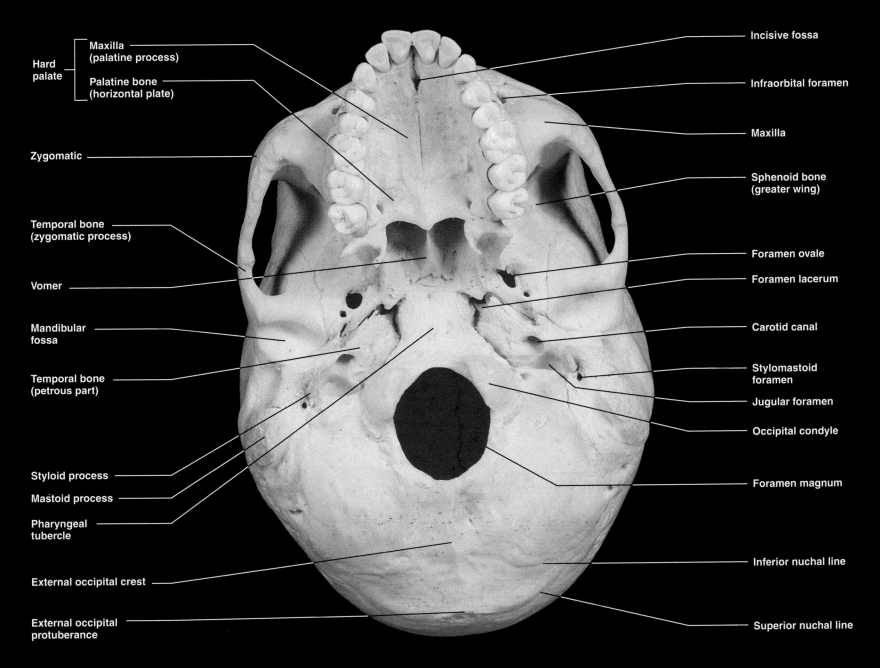

Figure 4 Skull, external view of base.

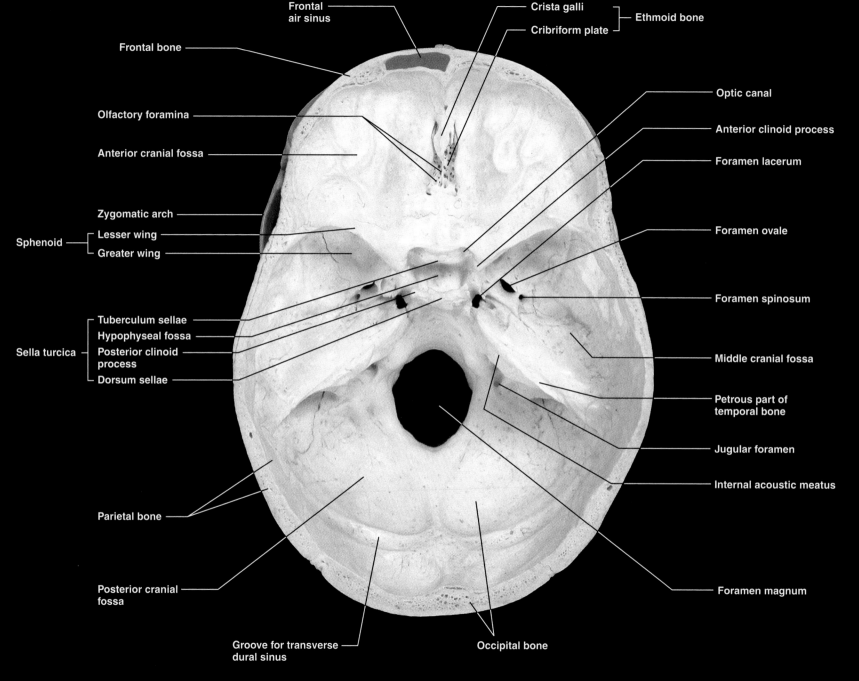

Figure 5 Skull, internal view of base.

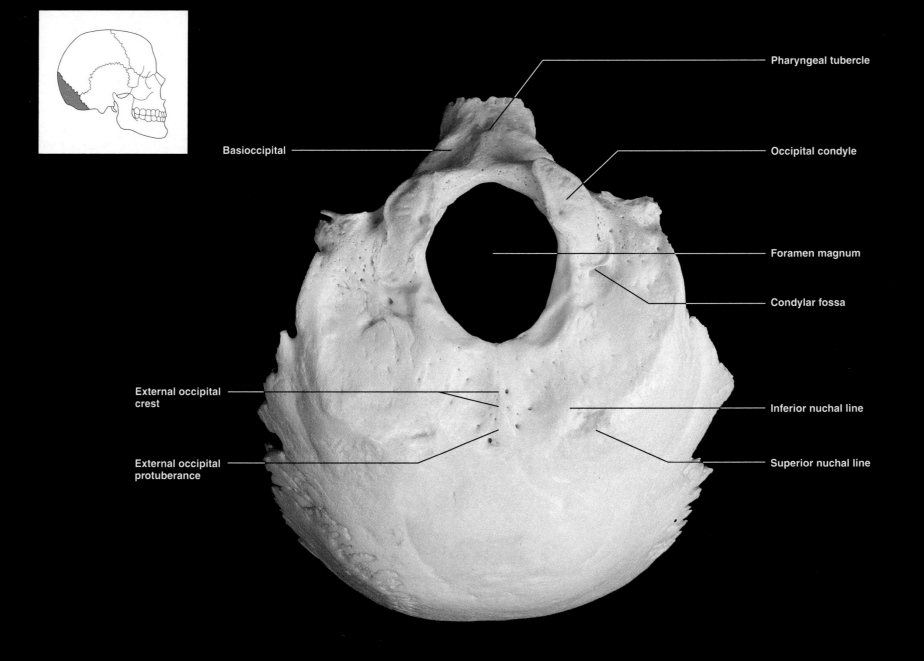

Figure 6 Occipital bone, inferior external view.

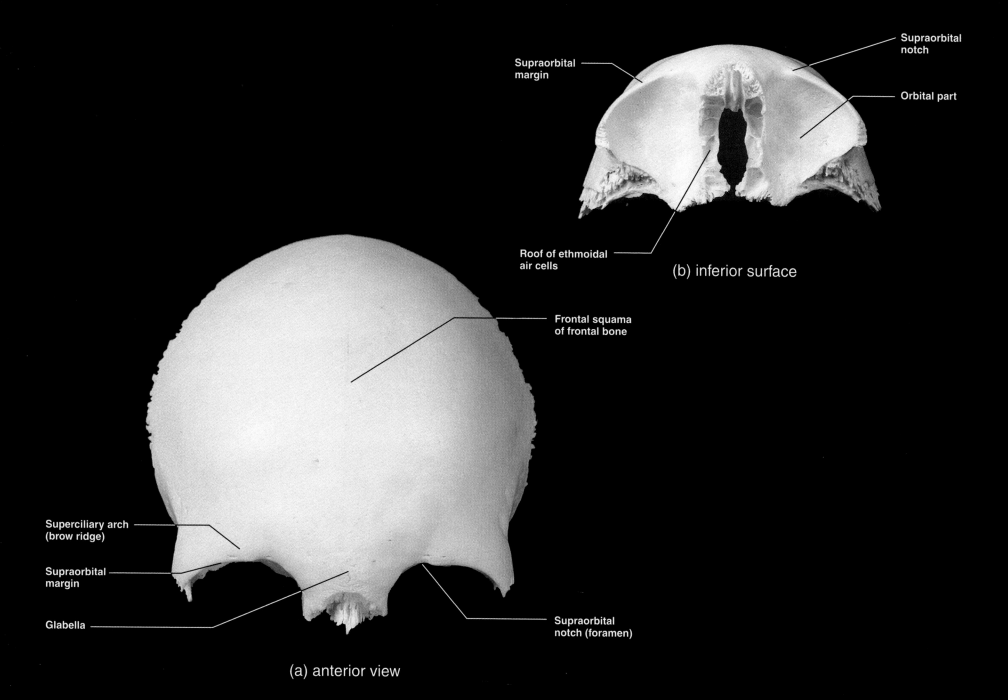

Figure 7 Frontal bone.

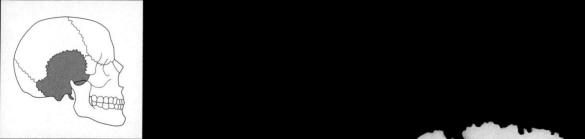

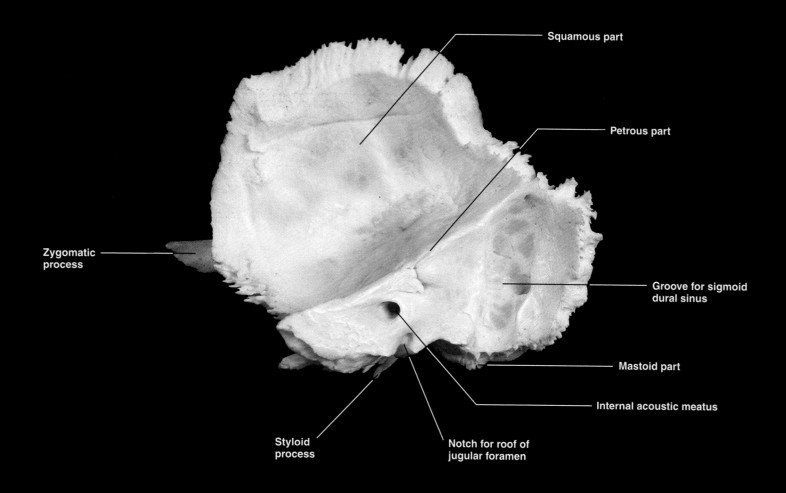

(b) right medial view

Figure 8 Temporal bone.

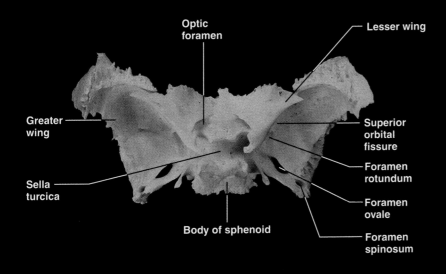
(a) superior view

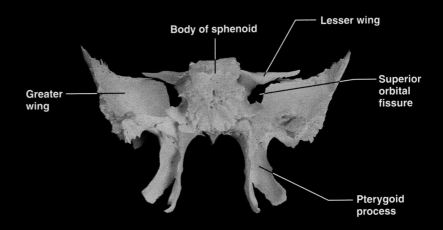
(b) posterior view

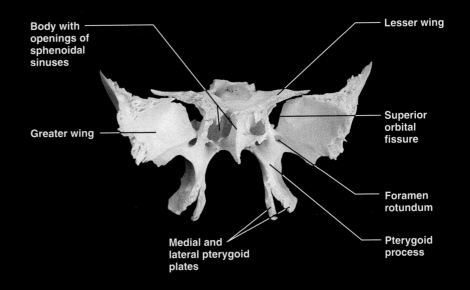
(c) anterior view

Figure 9 Sphenoid bone.

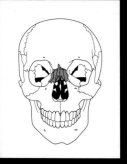
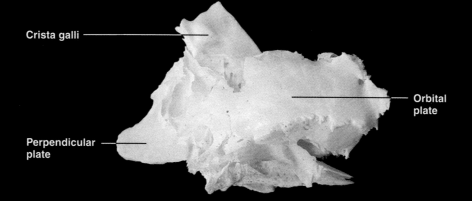
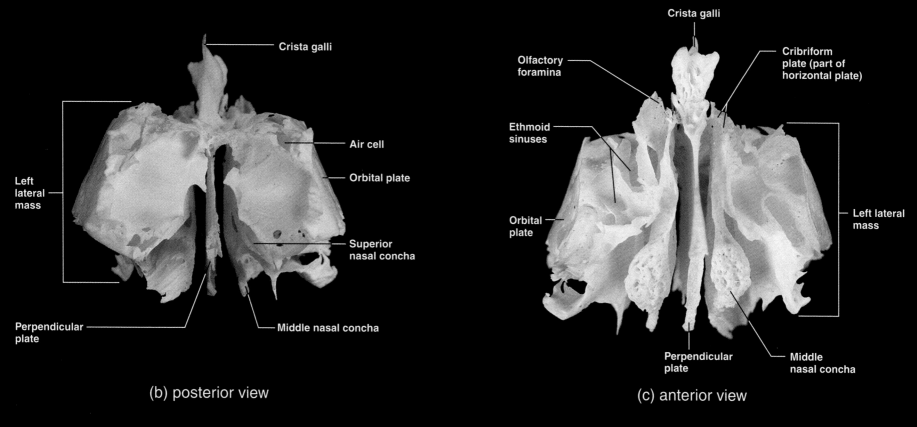

Figure 10 Ethmoid bone.

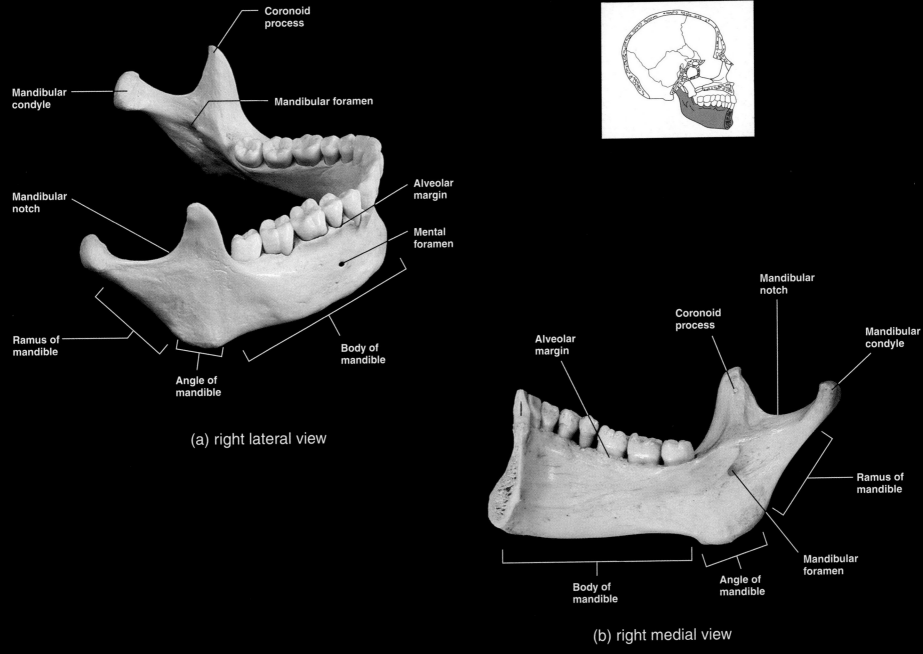

Figure 11 Mandible.

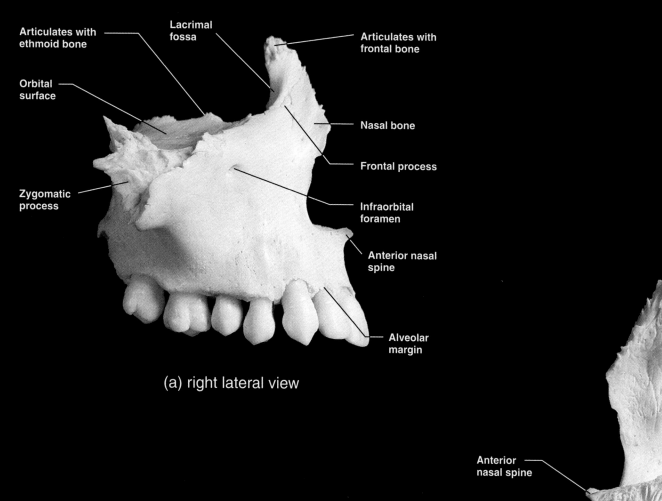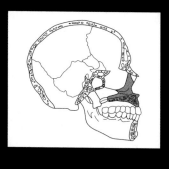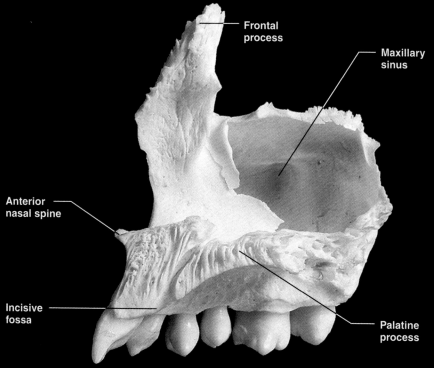

Figure 12 Maxilla.

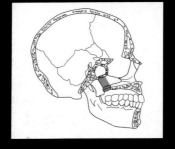

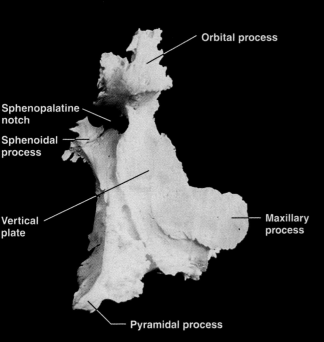

(a) right lateral view

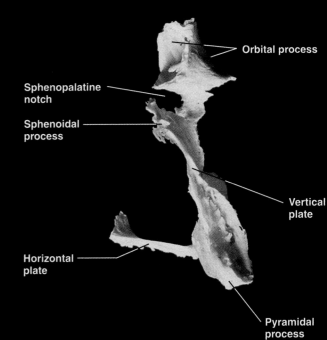

(b) right posterior view

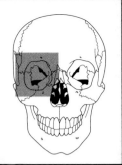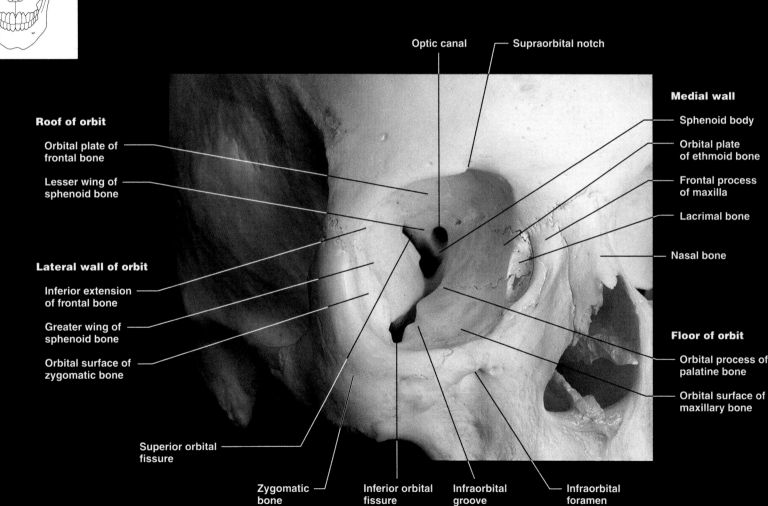

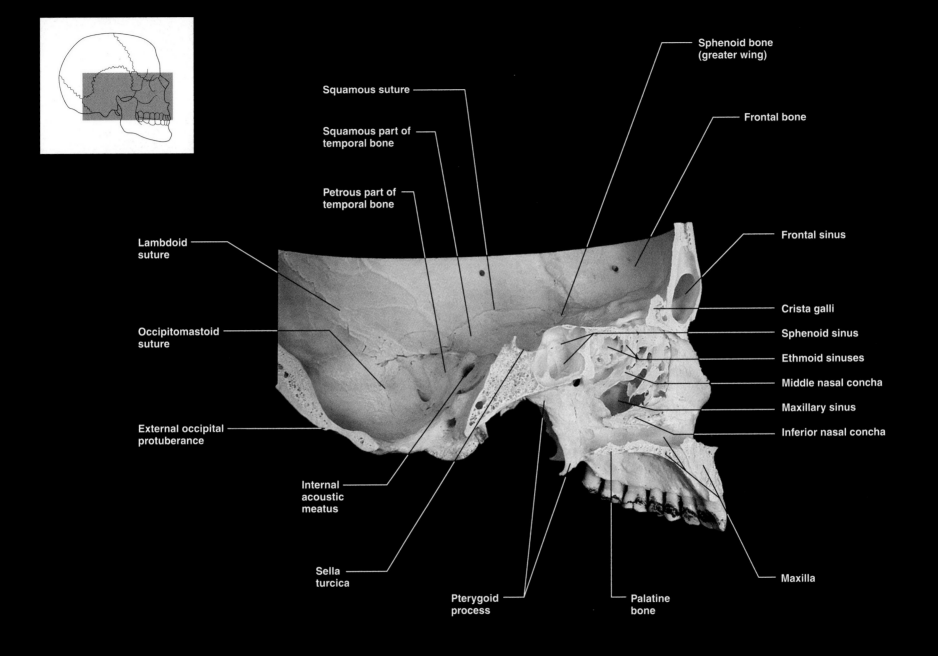

Figure 15 Nasal cavity, left lateral wall.

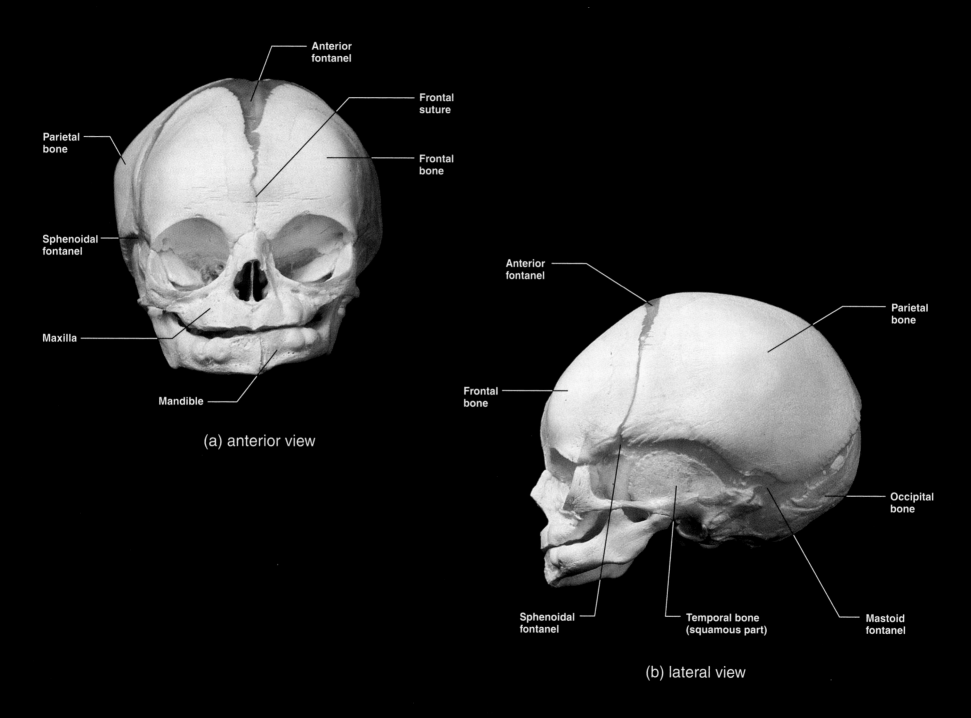

Figure 16 Fetal skull.

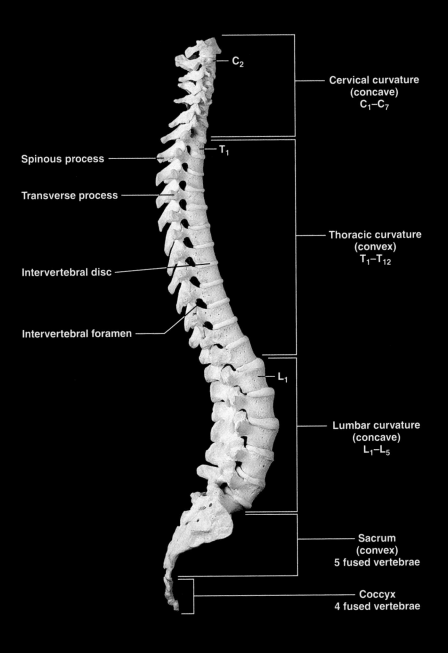

(a) right lateral view

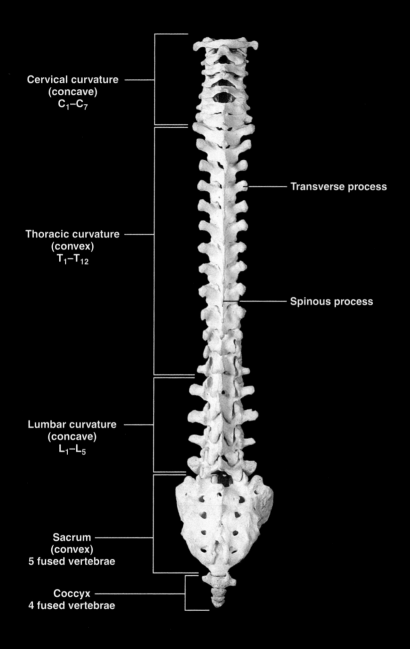

(b) posterior view

Figure 17 Articulated vertebral column.

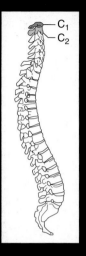
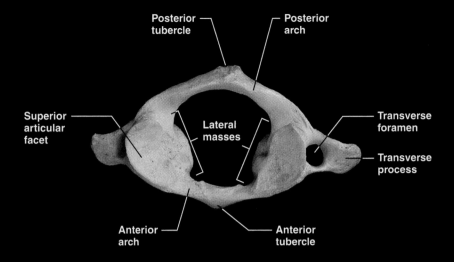

(a) atlas, superior view

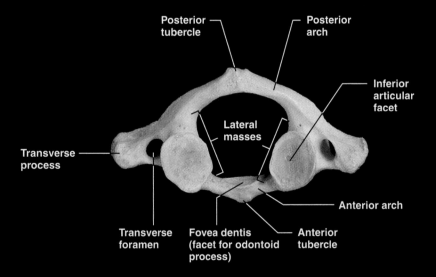

(b) atlas, inferior view

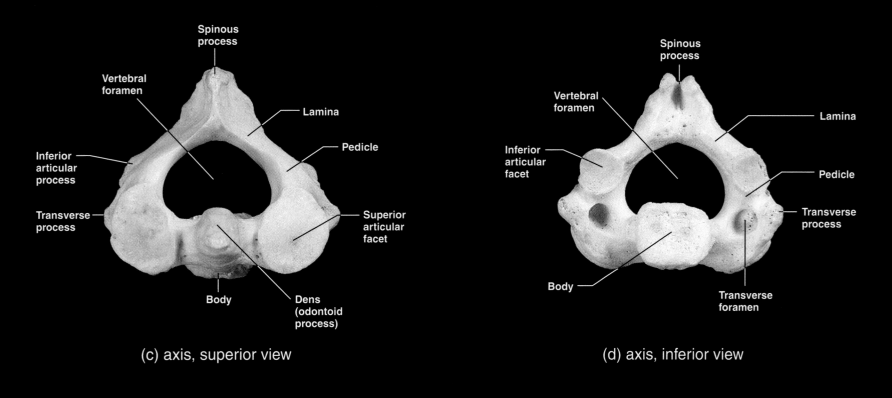
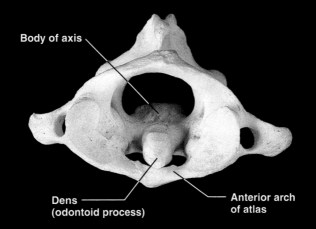

Figure 18 Various views of vertebrae C_1 and C_2.

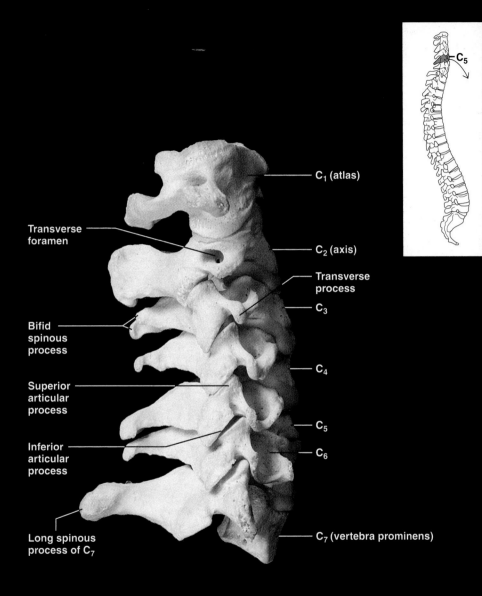

(a) right lateral view of articulated cervical vertebrae

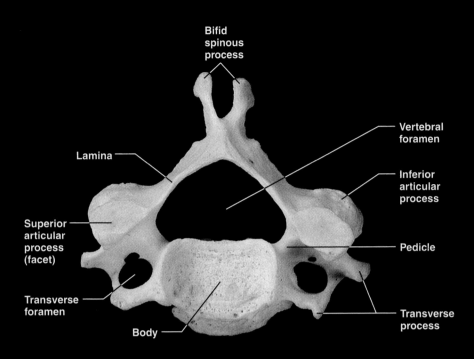

(b) fifth (typical) cervical vertebra, superior view

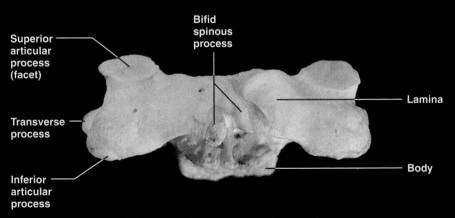

(c) fifth (typical) cervical vertebra, posterior view

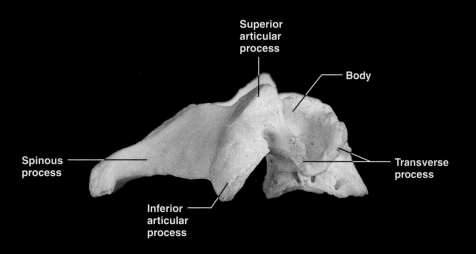

(d) fifth (typical) cervical vertebra, right lateral view

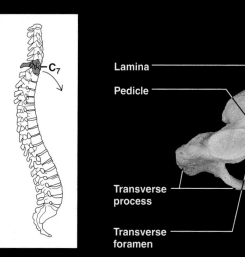

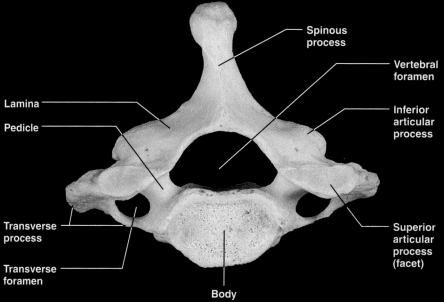

(e) vertebra prominens (C_7), superior view

Figure 19 Cervical vertebrae.

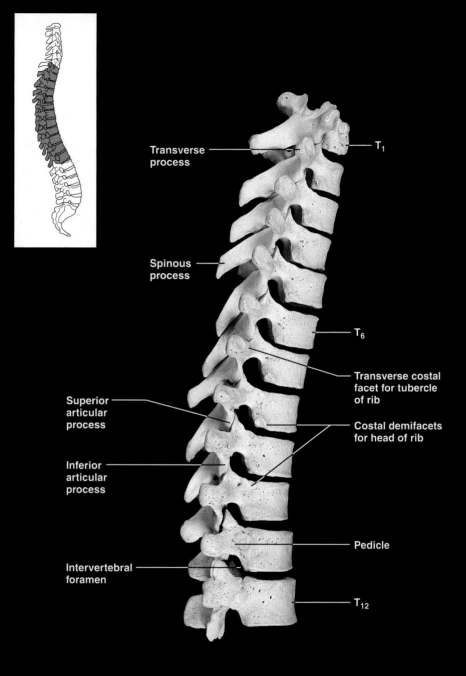

(a) articulated thoracic vertebrae, right lateral view

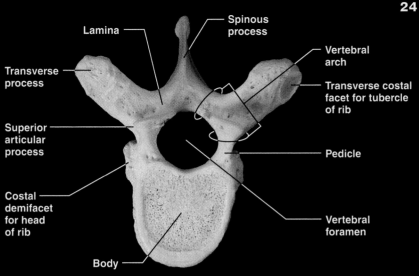

(b) seventh (typical) thoracic vertebra, superior view

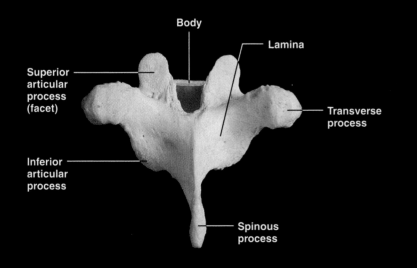

(c) seventh (typical) thoracic vertebra, posterior view

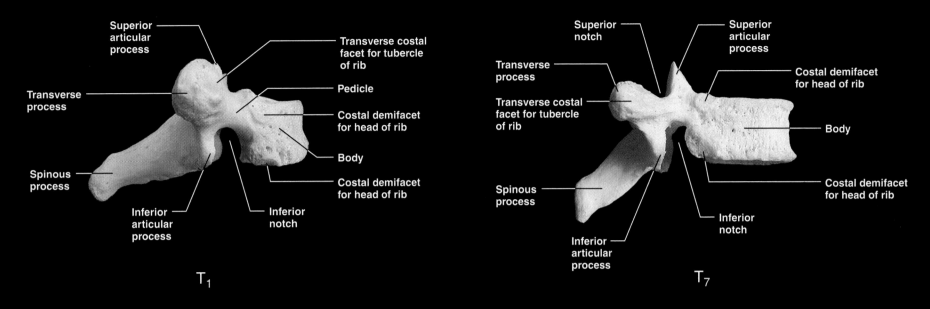
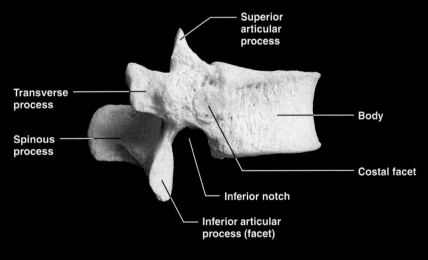

(d) comparison of T_1, T_7, and T_{12} in right lateral views

Figure 20 Thoracic vertebrae.

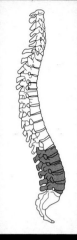
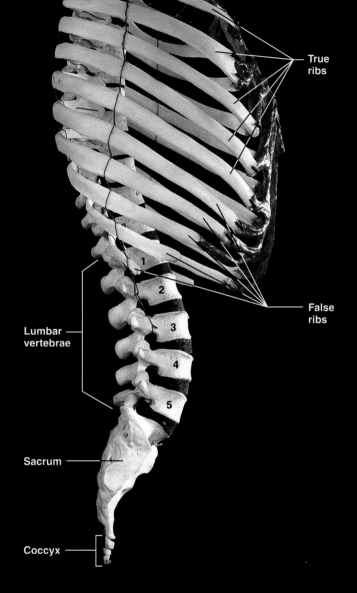

(a) articulated lumbar vertebrae and rib cage,

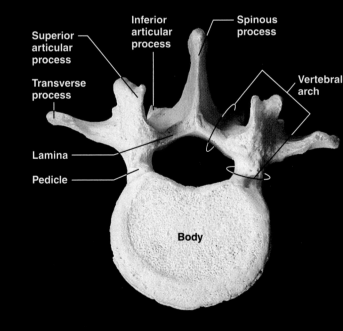

(b) second lumbar vertebra,

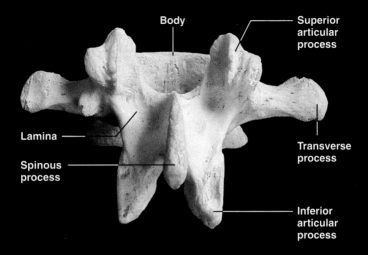

(c) second lumbar vertebra, posterior view

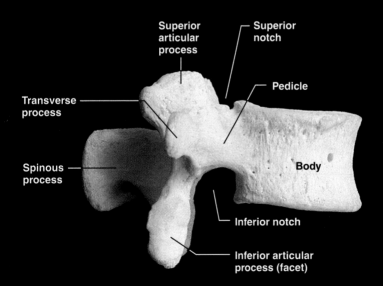

(d) second lumbar vertebra, right lateral view

Figure 21 Lumbar vertebrae.

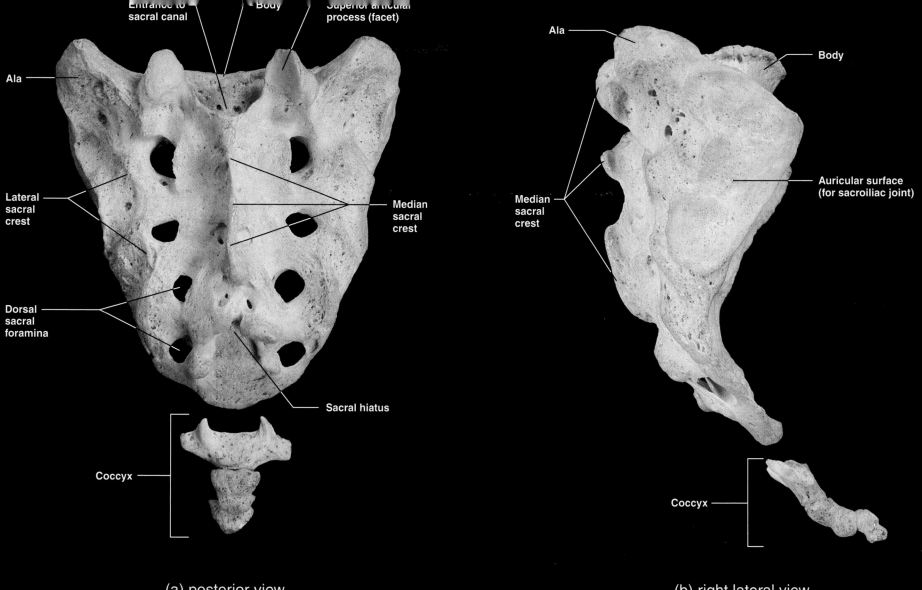

(a) posterior view

(b) right lateral view

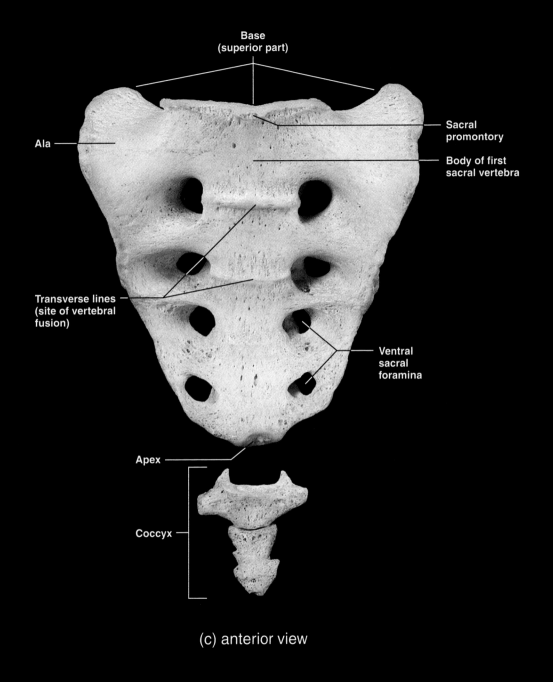

Figure 22 Sacrum and coccyx.

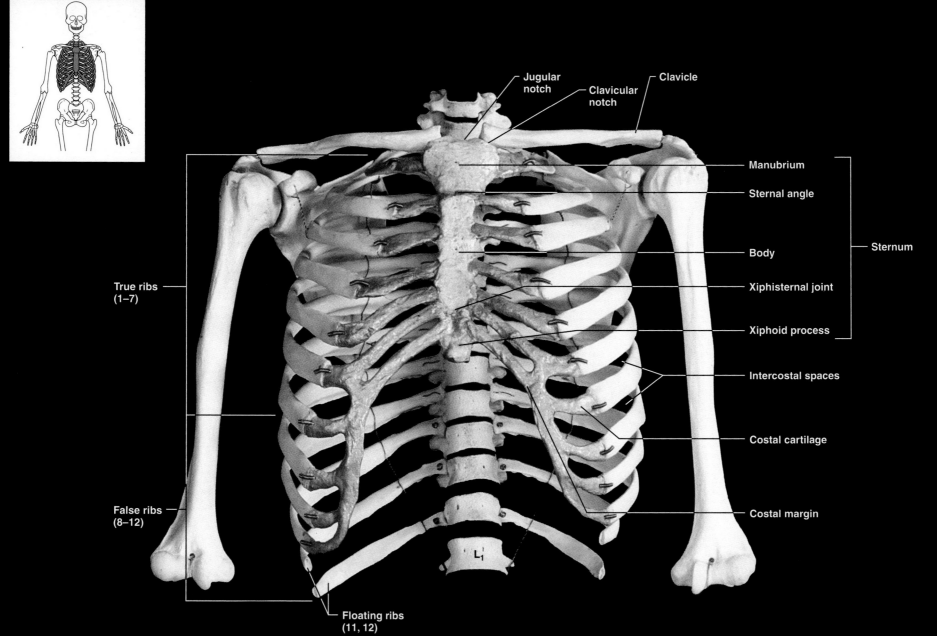

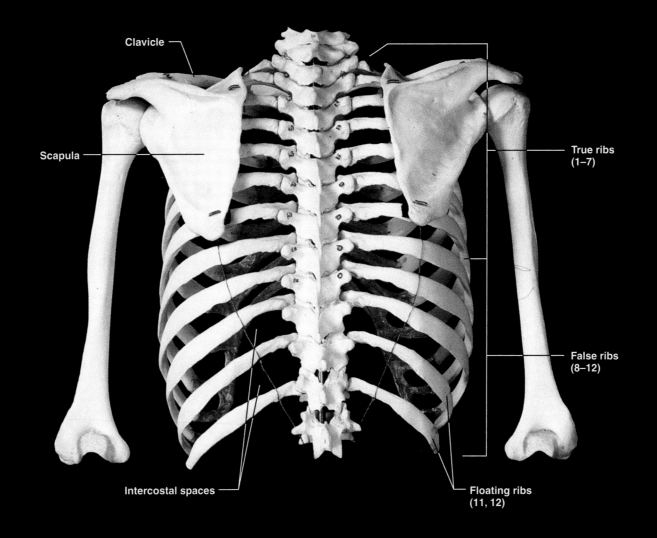

(b) posterior view

Figure 23 Bony thorax.

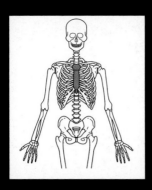
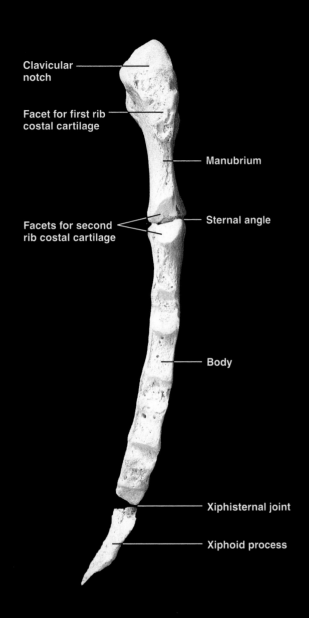
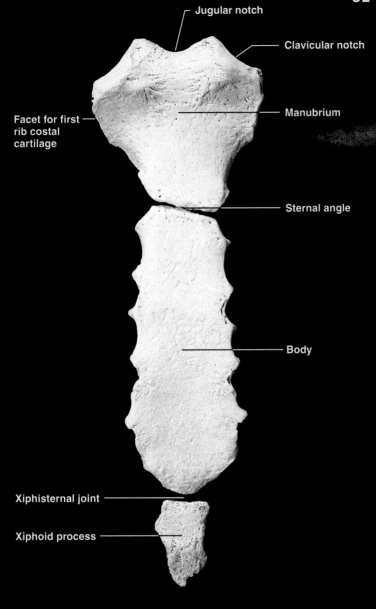

(c) sternum, right lateral view

(d) sternum, anterior view

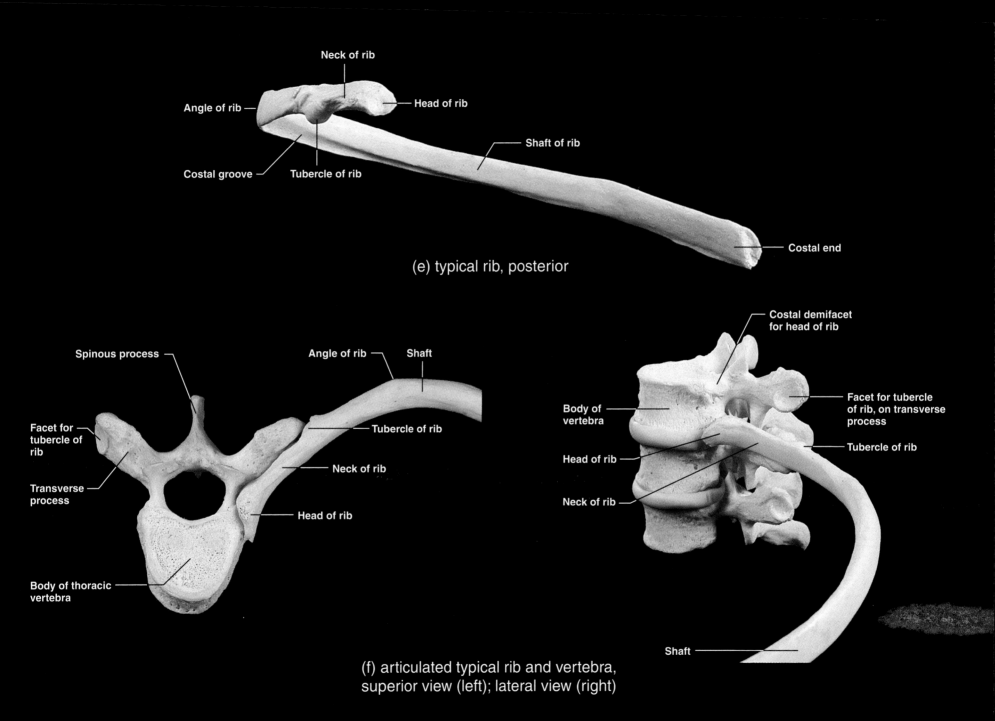

Figure 23 Bony thorax (continued).

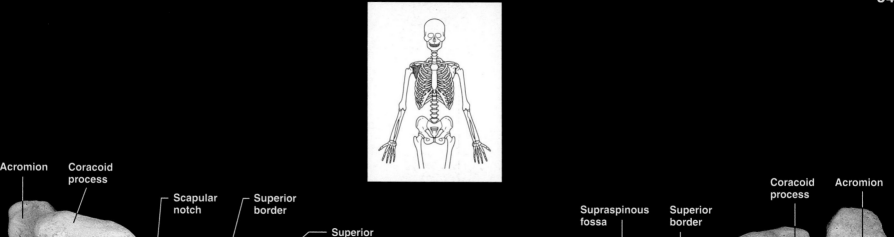

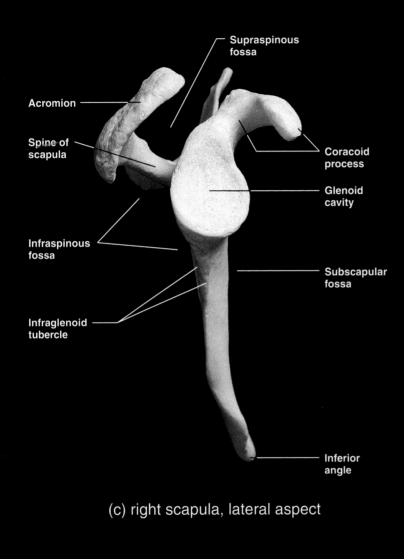

Figure 24 Scapula and clavicle.

(c) right scapula, lateral aspect

(d) right clavicle, inferior view (top) and superior view (bottom)

(e) articulated right clavicle and scapula, superior view

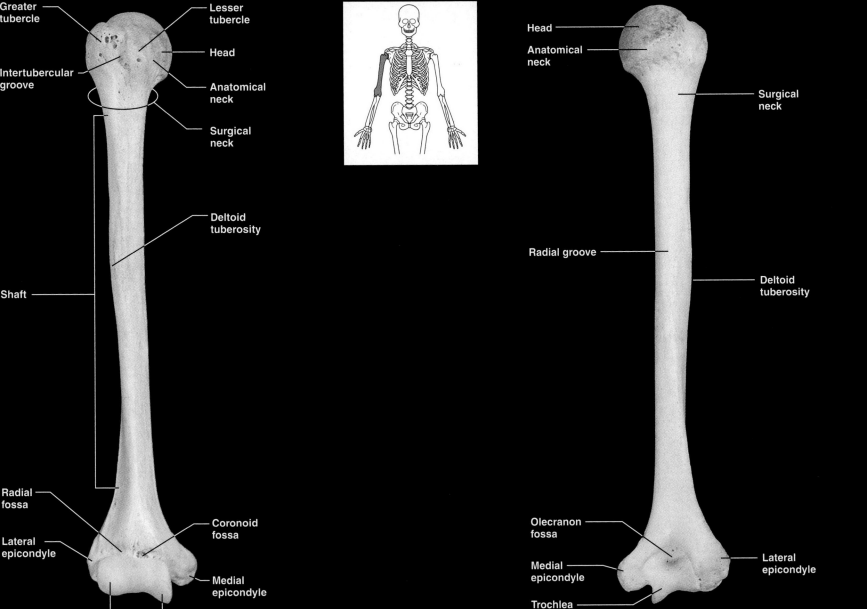

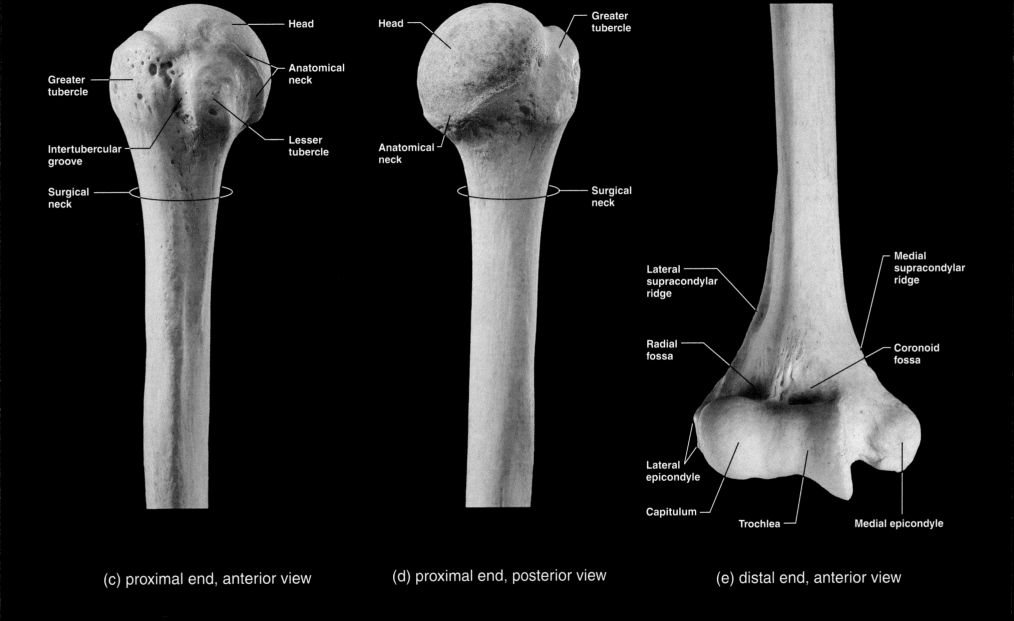

Figure 25 Right humerus.

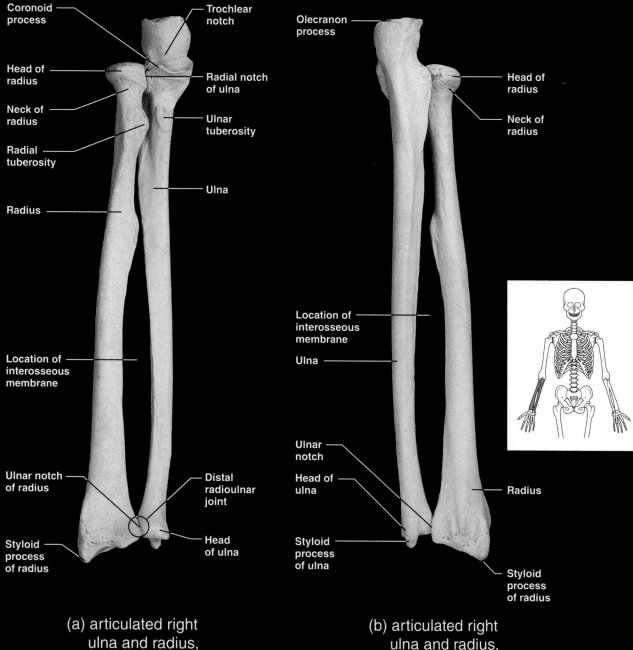
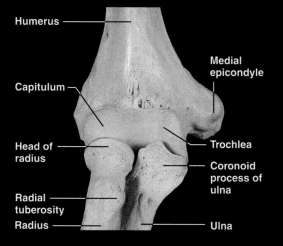
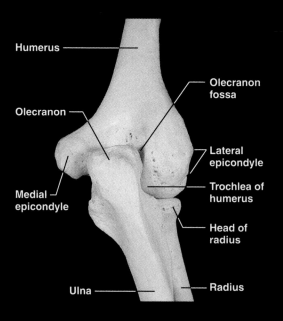

(a) articulated right ulna and radius, anterior view

(b) articulated right ulna and radius, posterior view

(c) articulated right humerus, ulna, and radius, anterior view

(d) articulated right humerus, ulna, and radius, posterior view

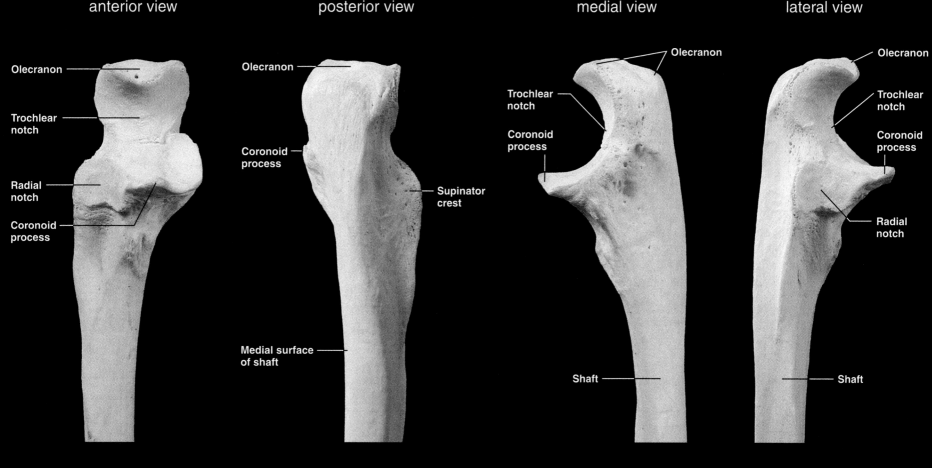

(e) right ulna, proximal end

Figure 26 Right ulna and radius.

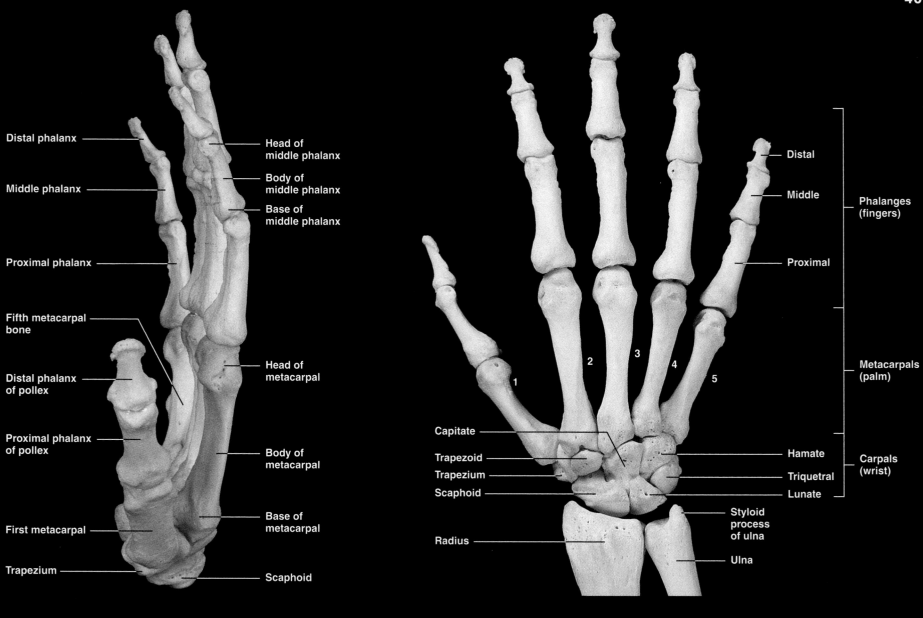

Figure 27 Bones of the right hand.

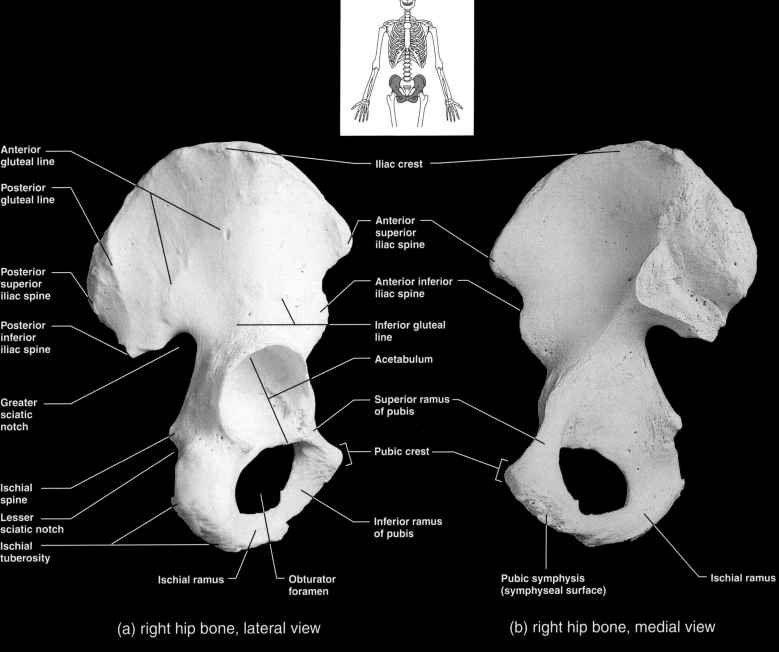

Figure 28 Bones of the male pelvis.

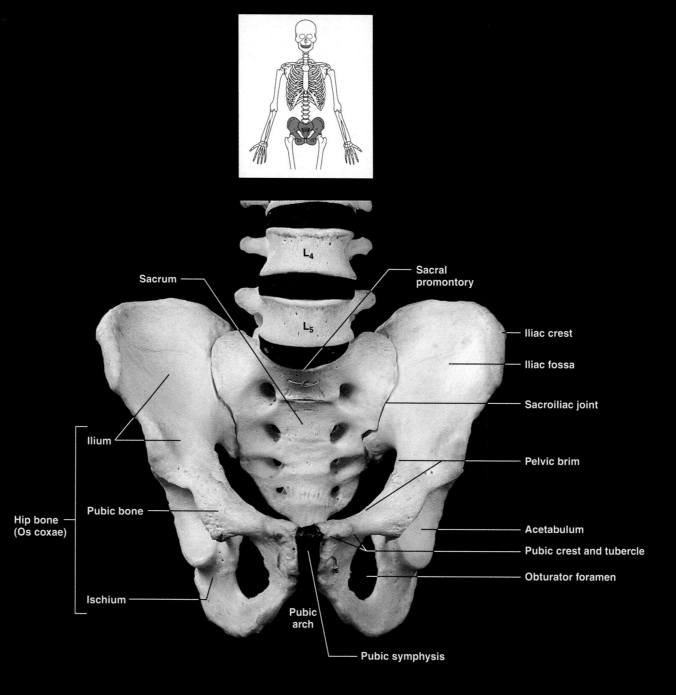

(c) articulated male pelvis, anterior view

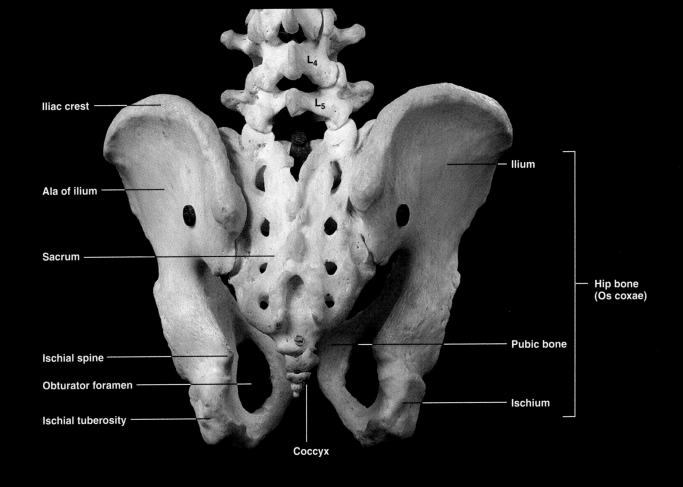

(d) articulated male pelvis, posterior view

Figure 28 Bones of the male pelvis (continued).

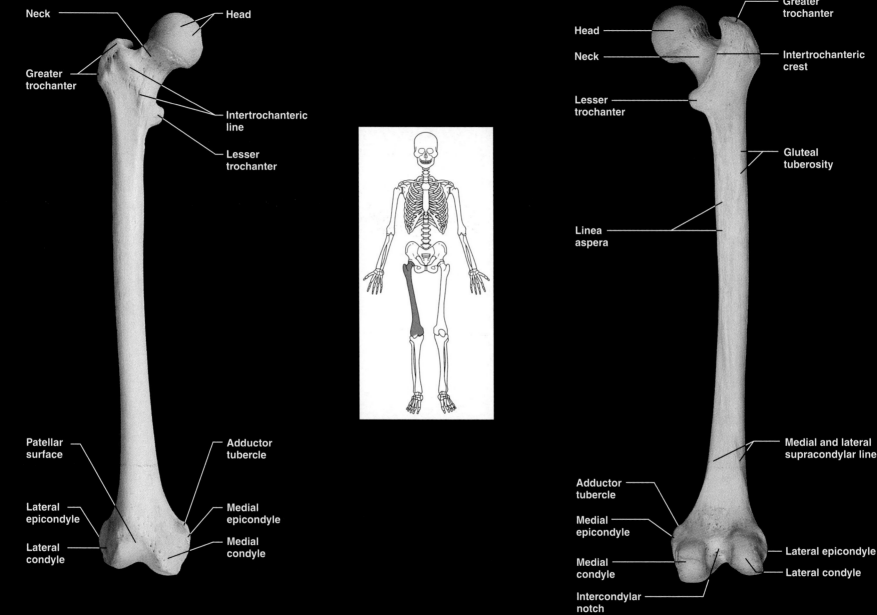

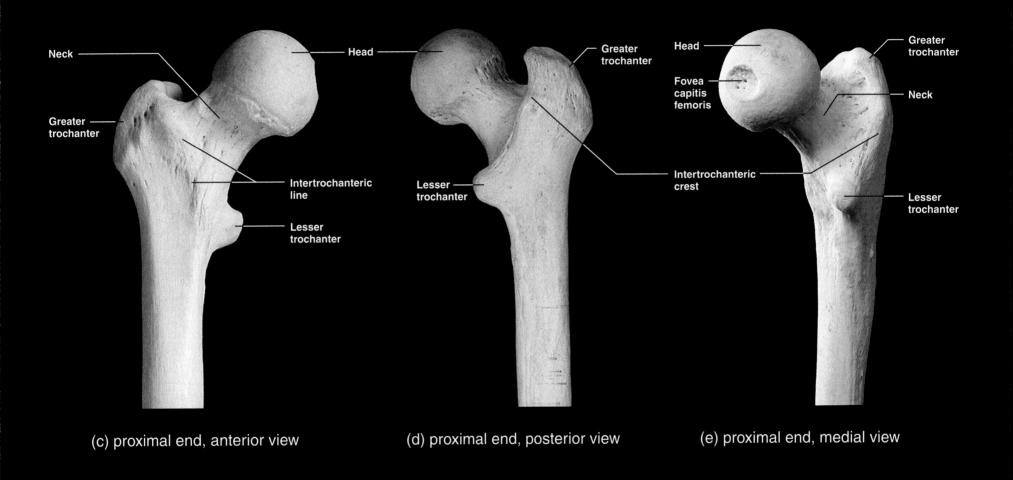

Figure 29 Right femur.

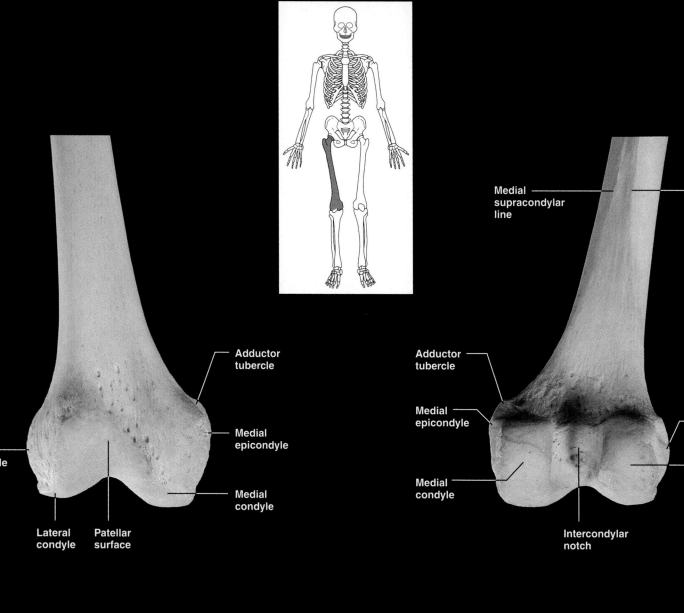

(f) distal end, anterior view

(g) distal end, posterior view

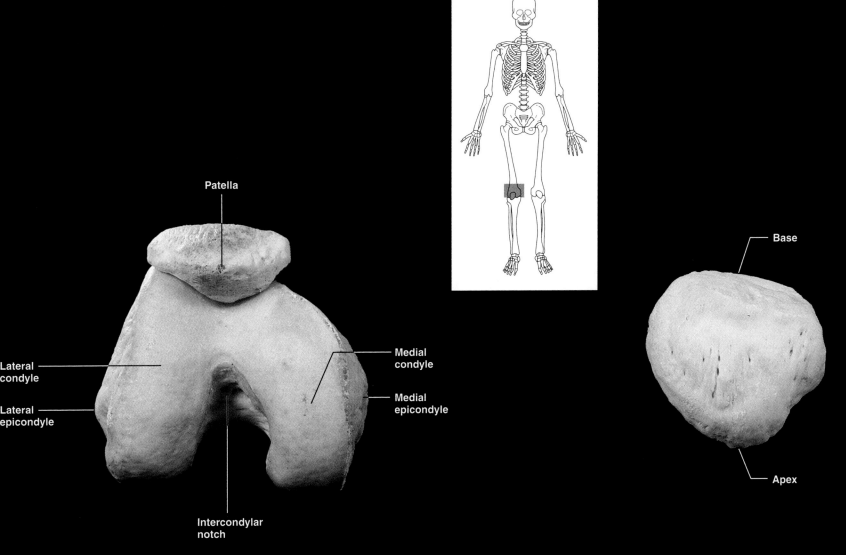

(h) Articulated right femur and patella, inferior view with knee extended

(i) Right patella, anterior surface

Figure 29 Right femur (continued).

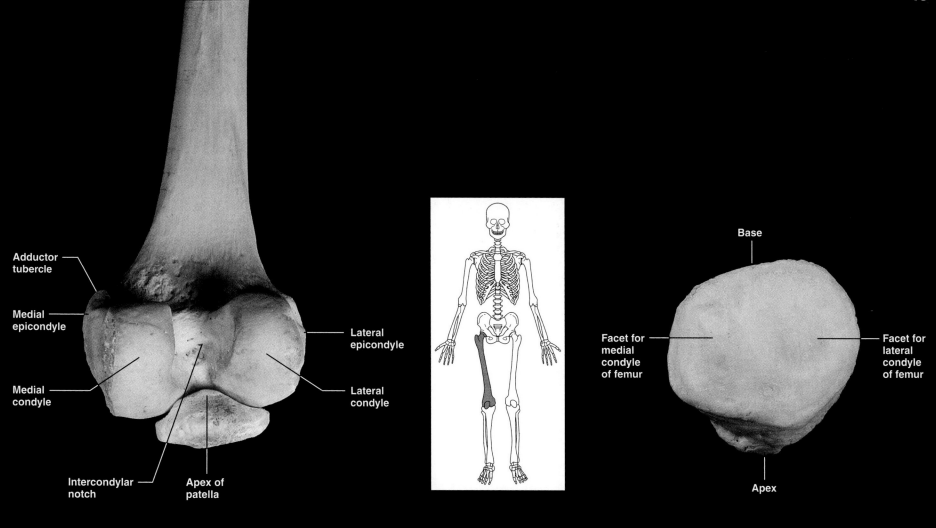

(j) Articulated right femur and patella, inferior posterior view with knee flexed

(k) Right patella, posterior surface

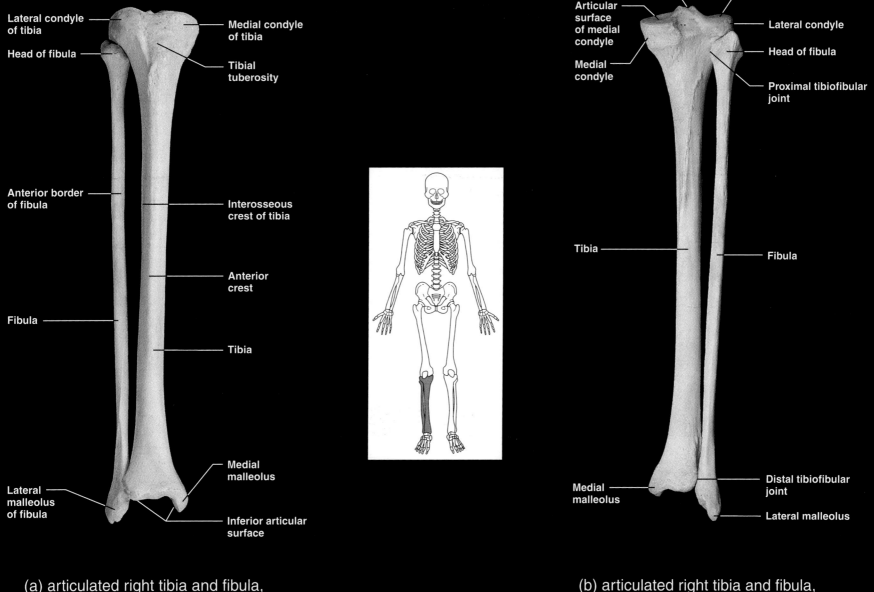

(a) articulated right tibia and fibula, anterior view

(b) articulated right tibia and fibula, posterior view

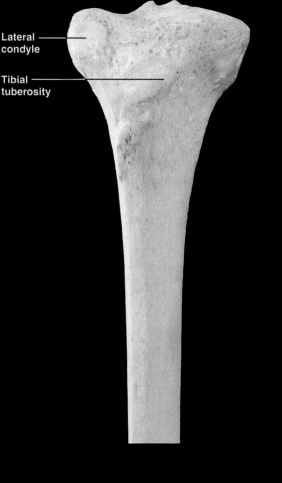

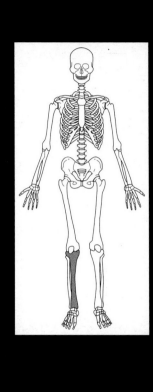

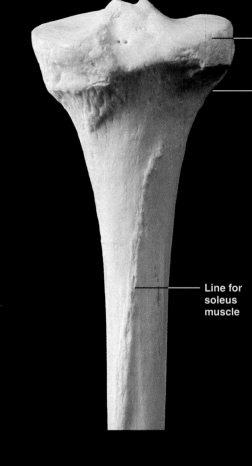

- Lateral condyle
- Tibial tuberosity

- Line for soleus muscle

(c) right tibia, proximal end, anterior view

(d) right tibia, proximal end, posterior view

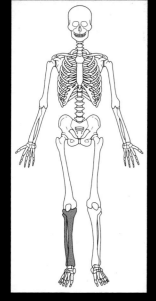
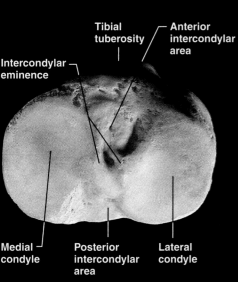
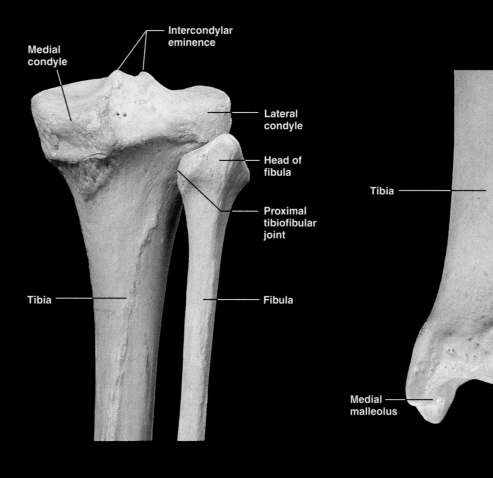

(e) right tibia, proximal end, articular surface

(f) articulated right tibia and fibula, proximal end, posterior view

(g) articulated right tibia and fibula, distal end, posterior view

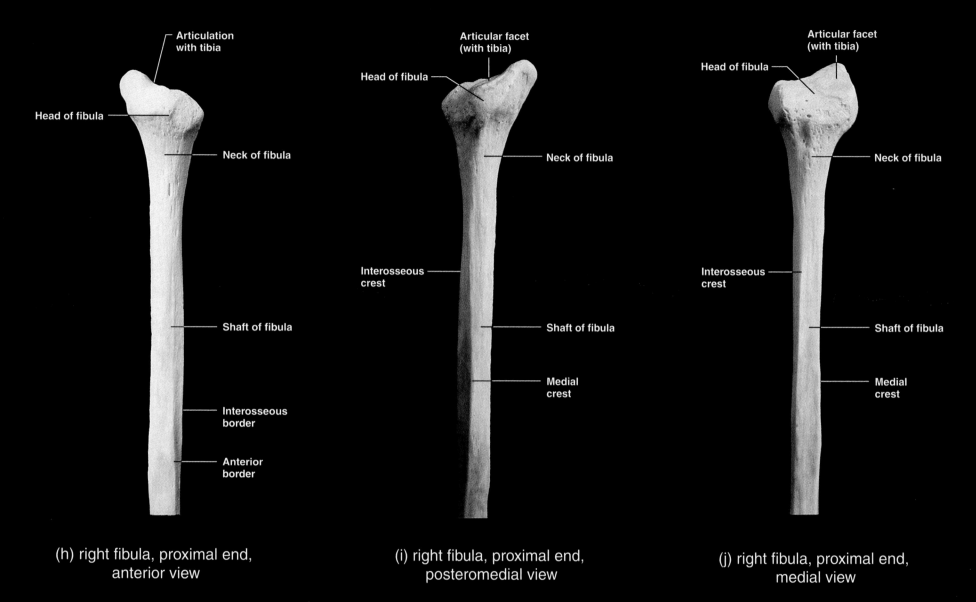

Figure 30 Right tibia and fibula (continued).

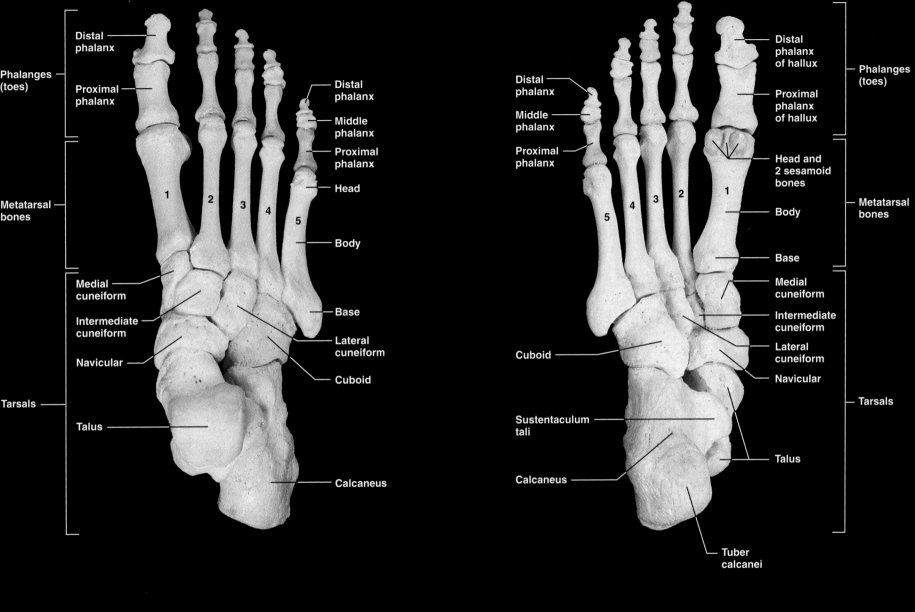

Figure 31 Bones of the right ankle and foot.

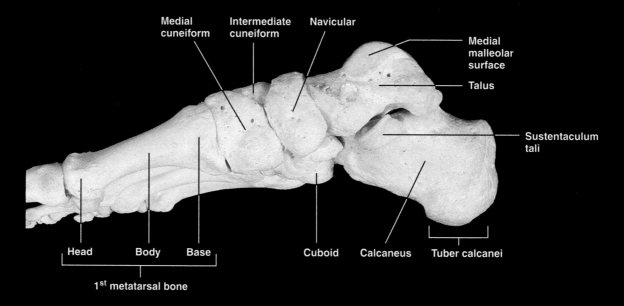

(c) medial view

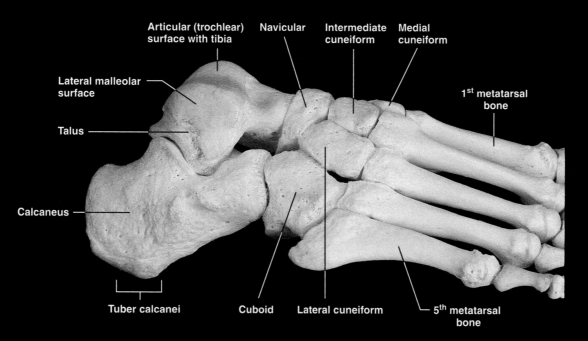

(d) lateral view

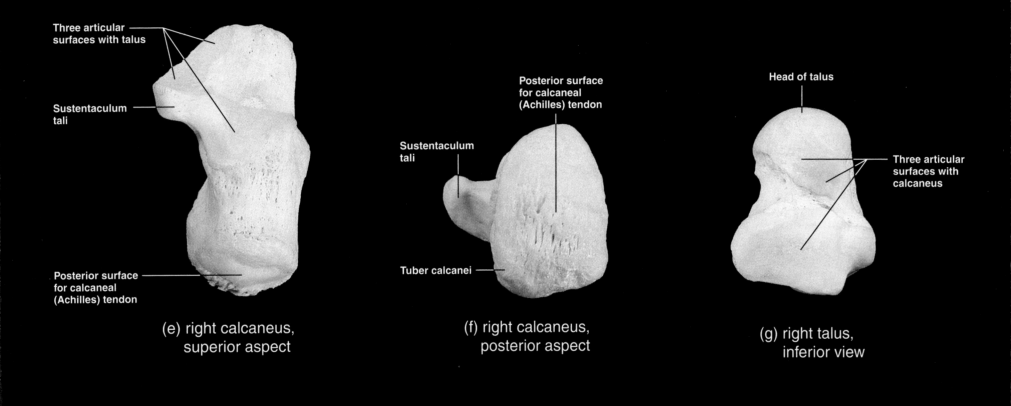

Figure 31 Bones and ankle of the right foot (continued).